The digitization of industry

Nassima Bouri

The digitization of industry

ScienciaScripts

Imprint

Any brand names and product names mentioned in this book are subject to trademark, brand or patent protection and are trademarks or registered trademarks of their respective holders. The use of brand names, product names, common names, trade names, product descriptions etc. even without a particular marking in this work is in no way to be construed to mean that such names may be regarded as unrestricted in respect of trademark and brand protection legislation and could thus be used by anyone.

Cover image: www.ingimage.com

This book is a translation from the original published under ISBN 978-620-2-54803-8.

Publisher:
Sciencia Scripts
is a trademark of
International Book Market Service Ltd., member of OmniScriptum Publishing Group
17 Meldrum Street, Beau Bassin 71504, Mauritius

ISBN: 978-620-3-13975-4

- *General Introduction*

The digital revolution in industry is the result of the convergence of several economic, industrial and technological trends born with Internet : the massive dematerialization of a growing number of our socio-economic activities and the interconnection of everything with everything (objects, machines, people).

The digital economy refers both to businesses and individuals who use ICT in their daily activities and to the ICT industry, which includes manufacturers and service providers. It has been defined as "the network formed by suppliers and users of digital content and technologies used in everyday life. This ubiquitous content and technologies are essential to virtually every activity in our economy and society. They enable businesses to be innovative and productive, governments to deliver services, and citizens to interact and exchange information and knowledge.

Indeed, the digital age is transforming everything: the nature of markets and products, how to produce, how to pay and how to pay, the scale of capital to be exploited globally, and human capital needs.

---

[1] Government of Canada, Public Consultation Canada Digital 150, 2010, p. 4.

It also boosts productivity, exposing companies to new ideas, technologies, new business and management models, and creating new market access channels. And all this at relatively low costs. It's no exaggeration to predict that companies will rely more and more on artificial intelligence for basic routines and more complex tasks.

For digital technologies to have an impact on economic development, appropriate policies must be put in place to remove the barriers preventing emerging economies from fully engaging in the digital economy and maximizing the benefits, while minimizing the risks.

The powers of digital transformation, artificial intelligence, massive data, the Internet of Things, uberization, cell phone and blockchain technologies are on the way to a fourth industrial revolution. This evolution is essentially beneficial for the improvement of the standard of living, and of the various economic and societal activities. However, it can also have disruptive effects, particularly on labor markets.

It is essential for social cohesion and the sustainability of our social models and strong economic institutions that rapid technological change is effectively managed to maximize benefits and minimize negative impacts. This includes equipping these entities with the tools and capabilities needed to participate fully and effectively in this digital transformation.

Faced with these rapid changes and the frantic pace of digital technology adoption, as well as with the emergence of a new economic order, organizations, and especially industries, are seeking to transform themselves rapidly. This upheaval at the dawn of the fourth industrial revolution, which is rapidly transforming the landscape of the industrial sector, is often referred to as Industry 4.0, intelligent factories, the Internet of Things, cyberphysical systems and digital transformation.

Industry 4.0 encompasses the digitization of horizontal and vertical value chains, product and service innovation and the creation of new business models. Key operational drivers of transformation include improving the customer experience, accelerating time to market and reducing costs. Indeed, the leaders of industrial companies who want to reap the benefits of this revolution are prioritizing Industry 4.0.

The creation of an Industry 4.0 oriented production environment is a phased approach that will take several years and will include the modernization of existing systems. Once the transformation begins, the opportunities for leveraging Industry 4.0 technology and concepts will be limitless. In this respect, this book will secondly analyze the economic and managerial impacts of these developments on the functioning of institutions, as well as their influences on the changing world of work in the last two chapters, and discuss policy options to address these challenges.

CHAPTER 1 :

Preamble on
The industrial economy

- *Chapter 1*: *Preamble on Industrial Economics*

- **Introduction to the Preliminary Chapter**

Indeed, the definition of industry refers to manufacturing production activities, it also refers to activities of transformation of raw materials or intermediate goods into products and services. An industry generally corresponds to the set of firms that produce closely related, closely substitutable goods or services, which therefore compete on the same market (Bouayad. Brahim, [2011]).

"This notion "Industry" comes from the term "*industria*" composed of "*indo*": *in* and "*struere*": to *build,* which means: the ability to do something, invention, know-how1 and, by extension, a profession that one exercises in order to live (mechanical, artistic or mercantile profession2).
The word took on a more restricted meaning in the 18th century, perhaps in Law'stime to refer to "all productive activity", i.e. all those that contribute to the production of wealth: the agricultural industry, the commercial industry and the manufacturing industry. Since the nineteenthcentury, the activities of

---

1Le Dictionnaire de l'Académie française 1694, volume 1.
2 Dictionary of the French language (Literary). Volume 3. 1873

agriculture are excluded from the field of industry, which now refers to "all socio-economicactivities based on the processing of raw materials1 ".

The main objective of this preliminary chapter is to present the basic principles of industrial economics, which consists in analyzing the functioning of markets and the behavior of firms in these markets. Indeed, it presents a preamble on the study of the digitization of industry after a review of the theoretical and empirical history of industrial economics.

1. History and definitions of the industrial economy

1.1. Definition of industrial economy

The main purpose of industrial economics is the study and analysis of the link between the strategies of firms and the evolution of different market structures (CPP, monopoly, oligopoly). Industrial economics is the branch of the economy that studies the functioning of markets and the behaviour of firms in these markets. It represents a field of economics devoted to understanding the functioning of a market according to its structure.

In particular, it deals with situations in which firms have market power, what economists call imperfect competition. In other words, industrial economics is concerned with the control of markets and the internal organization of companies. Because of its mathematical side, it is more oriented towards a "market control" approach.

---

[1] *"Industry:Etymology ofIndustry"*, on Centre National de Ressources Textuelles et Lexicales.

theoretical. This approach is thus apprehended by business school students.

The industrial economy depends on many variables describing the market, such as the number of sellers or the degree of vertical integration (i.e., whether the firm producing the good under study also owns the firm supplying the intermediate goods or the firm distributing the good).

According to this structure, the company's strategy will be analyzed in terms of price and quantity, but also in terms of quality, discrimination, R&D spending, advertising or innovation.

- Industrial economics is a field of economic analysis that aims to :
✓ To explain the functioning of the exchange relations between the producing companies which operate on the same market according to the different structures;
✓ Analyze the impact of these economic relations on the organization and functioning of the industry or market;
✓ Propose industrial policy methods and tools to industrial organizations, public authorities and regulatory authorities.

1.2. The history of the industrial economy

Historically, the term industrial economics was approved as a discipline in the 1940s by economists Edward Mason and Joe Bain. Starting in the 1970s, models in industrial economics were used to provide theoretical foundations for the development of the term.

which are based on game theory using modeling.

In France, the industrial organization has another name which is the industrial economy. The concept has had a slight delay in development in France compared to the United States. The industrial economy takes into account the success of markets in terms of efficiency.

To this end, two "traditions" of industrial economy1 oppose and complement each other historically:

1.2.1. The Harvard tradition

The first, called the Harvard tradition, dates from the 1920s and is mainly empirical. It developed around a "structure _ process _ performance" model.

The market structure (number of vendors, degree of product differentiation, cost structure, degree of vertical integration, etc.) is also important.) defines the processes (price, quality, R&D, investment, advertising,...) which will themselves define the market performance (efficiency, innovation, profit,...).

The emergence of the industrial economy is contemporary to that of the major industrial groups of the early 20th century, but for a long time remained an isolated field of empirical study without giving rise to economic theorizing or sophisticated econometric work.

---

[1] Renaud Bourlès, " Economie industrielle ", EAO-33-O-STRA, Central, Marseille, France, P 5, 2017.

1.2.2. The tradition of Chicago

Thus a new methodology has been developed since the 1970s. It is called the "Chicago tradition". This tradition is based on the need for a rigorous theory analyzing the various causal links related to the industrial economy. It then uses more empirical studies to identify the different competing theories.

Starting in the early 1980s, the adoption of gametheory tools to model corporate behaviour gave rise to a largely theoretical literature, produced and read by researchers who were previously non-specialists in industrial economics. In 1988, RichardSchmalensee defined industrial economics by three essential themes:

- The study of the determinants of the behaviour, size, scale and organization of private firms ;
- Imperfectcompetition, i.e. the extent to which the functioning and performance of the market. This theme covers in particular issues of price choice, quantity and capacity, as well as non-price competition: product selection, advertising, technical change.
- The study of publicpolicies concerning economic activity, in particular competitionlaw, deregulation, and privatization, as well as industrial policies affecting technical progress.

2. The presentation of the place of industrial economics in economic theory

The 1980s and early 1990s corresponded to the golden age of industrial economy and innovation, in the diversity of approaches developed. Several factors explain why it will at this time take a dominant position in the entire field of economic science and participate in its renewal.

- Firstly, it is a producer of innovative concepts and new approaches (innovation economy, theory of transaction costs, theory of perfectly contestable markets, knowledge economy, etc.).
- Secondly, it is in touch with the problems then facing economies in terms of competition through innovation or industrial reconversion (return to Schumpeter and the development of evolutionism): these are productive problems.
- Thirdly, the crisis of Keynesianism gave way to neoclassical approaches in macroeconomics and heterodox currents found in meso-economics a space and tools to think in their own field (until the return of institutionalism).

The approaches initiated in the field of industrial economics and innovation will find applications in new disciplinary fields which they will help to renew: banking economics, new international economy, spatial economics and, later on, geographical economics, development economics, etc. We can even speak of a trivialization as it is nowadays commonly accepted that competitive rivalry or competitive advantage depends on the ability of firms and economies to innovate.

- **Conclusion of the Preliminary Chapter**

The main purpose of industrial economics studies is to provide this characterization by using the schema that links market structures and corporate behavior. Jacquemin. A, 1989 drew two limits to this research, one at the theoretical level and the other at the empirical level. On the theoretical level, the concern to situate the analysis in the context of a precise microeconomic model was most often absent and the type of oligopolistic interdependence was rarely explained.

Thus on an empirical level, two types of studies characterize the traditional optics. Firstly, case studies, particularly prolific in the 1960s, provided in-depth knowledge of certain industries such as steel, oil and automobiles. The consideration of qualitative aspects shed light on the complexity of industrial reality, while quantitative measures, such as the degree of concentration or profit rates, provided simple synthetic indicators of the observed situation" (Jacquemin. Alexis, [1989]).

On the other hand, the new industrial economy was openly different from pure and perfect competition, which is not a realistic and relevant description of the markets. The model of pure and perfect competition is based on the following assumptions:

✓ An atomicity of supply and demand, which translates into a lack of market power of sellers and buyers ;

✓ A homogeneity of the goods offered on the market, which translates into a perfect substitutability of the sellers;

✓ Free market entry for sellers and buyers and free movement of factors of production;

✓ Perfectly informed buyers and sellers.

In conclusion, it should be added that the industrial economy offers tools and models to decipher the competitive strategies of companies. The aim is to better understand how companies interact with their competitors, but also with their customers, suppliers or regulatory authorities.

CHAPTER 2 :

Industry 4.0 :

Automation, programming and robotics

- *__Chapter 2__ : The industry4.0 :   Automation, programming and robotization*

- **Introduction to Chapter 2**

Industry 4.0 promotes what is called a "smart factory". In structured modular intelligent factories, cyber-physical systems monitor physical processes, create a virtual copy of the physical world, and make decentralized decisions. On the Internet of Things, cyber-physical systems communicate and cooperate with each other and with humans in real time, both internally and through the organizational services offered and used by participants in the value chain.

The concept of Industry 4.0 or Industry of the Future is a new way of organizing the means of production. This new industry asserts itself as the convergence of the virtualworld, digital design, management (finance and marketing) with the products and objects of the real world.

The great promise of this fourth industrialrevolution is to seduce consumers with unique and personalized products, and despite low manufacturing volumes, to maintain gains.

This industrialrevolution is influencing different aspects of our modern societies as a whole and the support of employees on the other hand. New challenges are emerging through this new way of producing. Industry 4.0 obviously affects the economic aspect but also has social, political or environmental impacts. It raises the question of the employment of millions of employees throughout the world.

1. Origin and definition of the concept

Industry 4.0 is the cyber-physical transformation of manufacturing. The name is inspired by German Industry 4.0, a government initiative to promote connected manufacturing and digital convergence between industry, business and other processes.

Industry 4.0 is the fourth industrial revolution, but there is disagreement on how to define revolutions. The first industrial revolution took place at the end of the 18th century and was marked by the mechanization made possible by steam and hydraulic power.

The second industrial revolution, which took place at the beginning of the 20th century, was driven by electricity and marked by mass production, assembly lines and the division of labor. The third, in the early 1970s, came with the use of computers to further automate machines and production processes.

The vision of the fourth industrial revolution will result in a smart factory and fully exploit digital manufacturing.

Currently in its infancy and beginning to unfold in isolation, its fully connected and vast form remains a vision for the future. It should be an end-to-end digitization of the manufacturing sector. In the vision of Industry 4.0, a fully interoperable ecosystem of machines and partners will be realized throughout the supply chain and the data will inform and correct the course of action.

2. Precursors of the concept

Industry 4.0 is one of the key projects of the Germangovernment's high tech strategy, which promotes the digitalrevolution in industry.

In France, companies such as Fives, Schneider Electric, Dassault Systèmes, Siemens, MecachromeAtlantique, sedApta-osys , Airbus Group, Bosch Rexroth, SNCF, are very involved in the development of Plant 4.0.

In recent years, France has seen the emergence of new startups such as Usitab, Optimistik and TellMePlus that exploit new technologies to improve and optimize French industry.

In July 2015, the Industry of the Future Alliance was created at the initiative of 11 industry and digital professionalorganizations, academic institutions (Arts & Métiers ParisTech, Institut Mines-Télécom) and technological institutions (CEA, CETIM) to federate initiatives to modernize and transform the industry in France.

ELCIMAI initiates a research program on the modular commonplace plant (real estate component) on the plant 4.0 project.

The Bosch Group, together with its Bosch Rexroth division, organized the "Tech Day Industrie 4.0" in France on November 18, 2015, a day of discussions and conferences on concrete solutions for the industry of the future. By sharing its vision and practices with French manufacturers, Bosch underlines its willingness to open up its know-how to better enable the exchange of ideas and foster partnerships in the key sector of industry.

In the U.S., the Coalition for Leadership in Intelligent Manufacturing project is also working on the future of industrial manufacturing. Independently, General Electric has been working for a few years on a project called the *Industrial Internet,* which seeks to combine the advances of two revolutions; the multiplicity of machines, devices and networks resulting from the industrial revolution, and the more recent developments in information and communication systems brought about by the Internet revolution.

3. Challenges 4.0

The implementation of a digital strategy integrated with the company's strategy, and including the acquisition of new technologies, represents certain challenges. Here are the main ones:

- Connectivity of software and hardware, even existing equipment.

- Standardization of standards and processes that facilitate data sharing.
- Re-engineering work methods and processes.
- Cybersecurity management to protect sensitive information and know-how.
- Access to specialists in digital technologies.
- The development of new skills.

4. Industry4.0, automation,programming, robotization: What links?

The development of the last decades has focused on controllers and their ability to communicate in a more user-friendly way. Communication protocols have evolved and the information gathered by them during operation can now even be shared with MES software, among others.

Standardization of communication is one of the issues of Industry 4.0, and manufacturers are devoting efforts to create standards in communication protocols.

The latest equipment is even connected to the corporate Ethernet network. This can be the first steps towards Industry 4.0, regardless of the size of the company's machine park.

4.1. Programmable Logic Controllers (PLC)

A programmable *logic controller* (PLC) is an electronic system for controlling logical, sequential and combinatorial processes in real time. These processes control production equipment, including servomotors, pumps and other processing equipment.

4.2. Robotics

The robot is a piece of equipment that needs programming to operate. This task used to require advanced computer skills. However, the constant evolution of technology has made programming increasingly user-friendly and intuitive. The robot can be manipulated manually by the operator in a job preparation mode, either using a simulation device or a teaching device.

4.3. Cobotics

Robotics is not new in itself, but its evolution has taken it to a new stage in terms of its possibilities of use. Collaborative robots, or "cobots", represent a major trend that reflects the new developments in robotics.

A cobot acts as an assistant and intervenes in a targeted way in complex and delicate tasks that cannot be automated. It also has learning features.

This type of robot is able to take an object, to give it to a human, in a cooperative environment that would not be adapted to more traditional robots. The cobots can be easily reprogrammed, moved (e.g. by mounting them on a mobile platform) and redeployed at different stages of the production line. The parts handled are generally smaller and lighter in weight than those handled by traditional robots.

- **Conclusion of Chapter 2**

Industry 4.0 is made possible by technologies that integrate both the digital and real worlds, for example :

✓ The Internet of Things (IoT): Connects an ever-increasing number of systems, devices, sensors, assets and people through both low-power wireless wide area networks and high-capacity wired networks;

✓ Mobile Solutions: Includes tablets and smart phones, handheld sensors and smart glasses.

✓ cloud computing: Includes cost-effective data processing and storage solutions;

✓ Cyberphysical Systems (CPS): Monitor and control physical processes using sensors, actuators and processors, based on numerical models of the physical environment - Massive Data Analysis and Business Intelligence: Transform data into actionable information, including early warning algorithms, predictive models, decision support tools, workflows and dashboards ;

✓ Advanced Manufacturing Technologies: Includes robotics and 3D printing.

Industry 4.0 directly follows the third industrial revolution, and started a few years ago. It is a kind of mix between all the latest technologies created.

CHAPTER 3 :

The transformation of the industry through digital technology

- **Introduction to Chapter 3**

The manufacturing industry is going through an extremely uncertain and unpredictable period with respect to consumer spending and confidence, and the geopolitical and broader macroeconomic situation is high. In the United States, the protectionist climate makes manufacturing one of the main objectives of the new presidency. In other regions, similar protectionist risks are present.

European manufacturers are mobilizing to advance the vision of Industry 4.0, despite the uncertainty that is growing even faster than before. Moreover, in other parts of the world, initiatives are being taken in an undeniable reality, where globalization has gone from an obvious fact for many to a source of mistrust for many others.

It is clear that under such conditions, automation and cost reduction are increasingly necessary, while increasing efficiency (increasing time to market, scanning and digitizing to maximize revenue, etc.).

It is most certainly here that we are seeing an even faster than expected adoption of the Internet of Things, in which the initial factors are the same as in the initial factors of digital transformation: increasing flexibility and reducing waste, reducing costs and improving efficiency, from manufacturing operations and business processes to maintenance and services.

1. The characteristics of a digitized factory

As we have already mentioned, Industry 4.0 is a concept that refers to a **fourth industrial revolution** and therefore, like the last three, to a new production method.

More concretely, the digitization of industry or factory 4.0 is an **interconnected system** that links machines, management methods (such as Enterprise Resource Planning, ERP) and products.

It is a "new workshop" that can take the following forms:

- ✓ An innovative plant
- ✓ A fully digitalized factory
- ✓ A flexible plant
- ✓ A factory focused on its external players
- ✓ A socially responsible factory
- ✓ A thrifty and environmentally responsible factory :

2. Issues around the digital industry

The industry of the future is a much more transversal notion, it responds to issues on 4 major levels:

- ✓ **Technological challenges:** markets are changing and the demands are increasing. It has never been more important to meet demand in a short time frame and at negotiated prices. At the same time, this same demand is taking off in customizable products, forcing industries to design processes that are agile and reconfigurable almost instantaneously.
- ✓ **Organizational issues:** machine agility also comes with organizational agility. Finally, the strategy of upgrading and innovation, coupled with

ever more flexible production, requires a rethink of management, particularly human resources management.

✓ **Environmental issues:** since the 1970s, the industry has been subject to legitimate environmental tensions. With resources becoming scarcer, climate change, and the energy transition gradually being initiated in all countries, Plant 4.0 needs more than ever to have a very small ecological footprint. This factory must also "think" about its products, which are often a major source of pollution and environmental debt.

✓ **Societal issues:** the industry must seek to reduce its negative impacts on its economic and social environment (negative externalities) and thus develop social responsibility (CSR). New technologies, as well as new relationships between the different stakeholders, new management methods, etc. are at the heart of Plant 4.0 and aim to respond to these challenges.

3. The Challenges of the Digital Transformation of the Industry (Industry Challenges 4.0)

3.1. Economic and strategic challenges

- The main challenges of the digitization of the manufacturing sector are as follows :
✓ The traditional challenge of skills shortages in the manufacturing sector is almost entirely related to the integration of computer and information technology (operational technology) and other technological and customer/service/innovation developments;

✓ An uncertain macro-economic and geopolitical context in which risk must be managed, cost reductions and increased efficiency inevitably ;

✓ A more complex and connected supply chain where data/information and speed are essential;

✓ The need to better understand the opportunities and benefits that can be achieved. While this is a strategic and informational issue, it also requires manufacturing companies to understand the technological drivers of new opportunities such as digital twins, robotics, artificial intelligence and 3D printing, among their benefits, use cases and overall context.

✓ An ever-changing customer, with a growing need to be not only more customer-centric, but also more adaptive and innovative. A highly competitive environment in which the fastest movers and shakers are on the verge of gaining advantages and even becoming disruptive ;

✓ The need to diversify and exploit new sources of revenue, by exploiting new ecosystems and (connected) data, to thrive and, in some cases, survive ;

✓ Lack of a clear vision and overall strategic approach to capitalize on the revenue growth and new revenue potential of Industry 4.0 ;

3.2. The challenges of the human factor

✓ Regarding the link between the digitization of Industry 4.0 and the adaptation of the "work L" factor, it should be pointed out that, the dimension of human talent in a changing reality where technology and innovation play deeper roles and talent in many of the areas mentioned (data, industrial IoT, convergence of information and telecommunications technologies, new business models, etc.), nor culture are present taking the necessary steps.

✓ In terms of the dimension of the talent and skills shortage, it is clear that as Industry 4.0 arrives and the digital transformation of manufacturing continues, the reality of work is changing.

✓ According to IDC[1] (end 2016 data), by 2020, 60% of workers in G2000 plants will be working alongside automated assistive technologies such as robotics, 3D printing, AI and AR / RV.

✓ In addition, automation, optimization and ongoing transformation have a human cost. From a purely commercial point of view, it is also a challenge.

✓ To this end, the human consequences must be addressed at a time of rapid digitization a threat.

---

[1] **IDC Manufacturing Insights:** Service Innovation and Connected Product Strategies The consulting service is ideally suited to meet the needs of companies: Manufacturers seeking technologies and business models for advanced diagnostics and services to improve product value and differentiation and increase service revenues.

Every organization, and certainly in the manufacturing sector, is a key element, it must be aware of the impact of automation and its role in society, as neglecting the human costs can lead to further erosion of brand equity and trust, as well as a decline in consumer confidence and purchases" Power.

4. Managerial strategies for the digital transformation of the industry

Maurice Ricci, [2016] has proposed a relevant process for the digital transformation of the industry based on several parameters:

4.1. Reconstruction of the company's business model

The integration of digital in the production process offers new opportunities for value creation to the company. It enables companies to rebuild the means not only to optimize their activity but also to reinvent themselves by positioning themselves differently on the value creation chain.

4.2. Innovation and accelerating the digital transition

In the face of technological change and the extent of the possibilities offered by scientific research, the capacity for innovation is becoming a key factor of differentiation, competitiveness and performance. It must become an essential axis of company development. Innovation is no longer the exclusive domain of technology and R&D. It invests the other dimensions of organisation : its processes, its way of working, its customer relations, its production line and, of course, its strategy. It becomes a process of transformation of the company as a whole and no longer just a way of improving products. It opens up to the ecosystem and takes advantage of it (Ricci1, [2016]).

Digital technology is an almost inexhaustible source of innovation both for the product, of course, but also for changing a use, a process, a business model. So what better way to start opening up than to get closer to the queens of secteur : digital start-ups and SMEs. By going to meet them, the company will familiarize itself with technologies and uses that it does not necessarily have in interne : connected products, 3D printing, mobile uses, big data, collaborative tools (Ricci, [2016]).

---

[1] Chairman of the Industry of the Future Committee Syntec Numérique.

4.3. Strengthening the Company/Customer relationship

The digital transformation of the industry is helping to optimize the relationship with the customer. Digital technology enriches the latter by bringing more interactivity, proximity and transparency. It facilitates a consistent customer experience across all interaction channels.

According to Ricci, [2016][1], collaborative tools and customer relationship management software (CRM - Customer Relationship Management, now called CX - Customer Experience) have this vocation. CX software, for example, provides a complete view of the customer and his situation (360° vision) by concentrating in a single space all the information related to it (marketing leads, opportunities, history, current contracts, after-sales data, commercial data, disputes, etc.); this, regardless of the channel used by the customer to contact the company (call center, email, mail, website, social media, etc.).

---

[1] According to the author, customer information must be accessible to all company departments in contact with the customer (marketing, sales department, sales administration, after-sales service, etc.) in order to ensure a coherent, efficient and fluid follow-up of the relationship. For his part, the customer lives a harmonized experience of his different modes of interaction with the company.

4.4. Acceleration of design

In a context of intense global competition, there is a constant need to innovate, reduce costs and time to market. To achieve this, companies have no choice but to design their products differently. This is especially true as the products themselves evolve and become more complex.

Digital offers a more open and collaborative approach to design. It democratizes 3D simulation and allows to better take into account the software in the product life cycle. It introduces new manufacturing techniques that broaden the horizon. In the digital world, the evolution of services and software tends to be continuous. To achieve this pace, digital companies are following two paths règles : they favor incrémentales developments; they conduct projects in an iterative and collaborative way (Ricci, [2016]).

4.5. The proactivity of the industrial chain

The digital answer to the question of production optimization is to drive it through data. How can this be done? By digitizing and interconnecting all the links in the value chain industrielle : from customer order to delivery, including procurement and interactions with suppliers (Ricci, [2016]).

Thus digitized and synchronized, the chain is characterized by its flexibility and modularity. It can be reconfigured automatically and adjusted according to demand. It adapts to inevitable variability (raw material prices and volumes, supply contingencies, machine breakdowns, quality fluctuations, etc.) while maintaining its quality objectives and optimized SRTs.

The digitization of the line begins with the digitization of the factory. It is obtained by connecting all its elements to the IoT constitutifs : machines, parts, products, workstations. Equipped with sensors and transmitters, these cyber-physical systems communicate with each other and interact continuously through networks, adaptation gateways and data exchange platforms.

4.6. The revaluation of the role of the human being

To restore the desire to work in the factory. Digital technology can contribute to this other ambition of the Industry of the Future program, by taking better account of the human aspects and by enhancing the role of the operator.

In a plant where machines are becoming more autonomous, the role of the operator is evolving towards that of a responsible pilot. Equipped with a connected mobile terminal (smartphone, tablet, etc.), possibly reinforced, he gains in autonomy, moves around and remotely monitors the operations in progress on several machines. Notifications on his mobile terminal alert him of incidents. He has instant access to the necessary documentation (Ricci, [2016]).

- Conclusion of Chapter 3

In the context of Industry 4.0, there are a number of security features that need to be considered. Production systems have important requirements in terms of reliability, availability and robustness. Breakdowns and disruptions must be avoided.

In addition, access to production-related data and services must be controllable to protect the company's know-how and prevent economic damage. Security is therefore essential to the success of intelligent manufacturing systems. It is important to ensure that production facilities do not present any danger to people or the environment, but also the data and information they contain. They must be protected against misuse and unauthorized access.

- *General Conclusion*

The digital economy, and the economy of the Internet and its commercial uses have aroused great interest. However, computer and networking technologies have the potential to transform not only consumer behaviour, but also the way the economy works. An economy characterized by these technologies is the digital economy, where market agents behave differently according to different economic rules than in the physical economy.

On the other hand, it should be added that, digitization has justified consequences on practically all cultural fields and in particular socio-economic. In connection with this new basic trend, new topics are gaining in importance related to data security, data protection, cybercrime, as well as business management in the digital age, training and further education, social media, the sharing economy etc.

We have tried through these integrated courses to discuss some of the economic implications of digital technologies.

We tried to present the main research questions on techniques and tools of the digital economy, the effects of the 4.0 industry, digital payment, and online economic transactions, artificial intelligence, techniques and pillars of Big Data, virtual currency policies, fiscal and monetary issues, and globalization of markets via GAFA and BATX.

Finally, it should be added that Industry 4.0 will change the face of the entire manufacturing system, from its architecture and organizational structure to products, services and business models. While the development and implementation of these solutions will be gradual and part of a long-term trend, the economic climate is already favourable. Firms that are not aware of new technologies and do not invest in pilot projects will lose their competitive advantage and miss the opportunity to lead the transformation that is currently sweeping the manufacturing sector.

Our objective was to demonstrate how fundamental are the economic transformations triggered by the use of information and communication technologies.

We conclude by noting at the international level that digital and computer and networking technologies not only improve the economic efficiency of financial institutions and businesses, but also present a new type of fictitious and digital market that could be an example of a perfect market requiring a more vigorous re-examination of the economic assumptions and results provided by studies conducted in traditional markets.

In addition, on the national level, Algeria has a significant development potential in the digital sector in the coming years, it represents however a weak dynamic in the digital field, which is indeed mainly due to a significant lack of support and funding, as well as a weak development of innovation activities. Finally, we hope that our country will be led to catch up a large part of its delay compared to emerging and developed countries in terms of added value achieved by ICTs.

Bibliographic
Sources

- *Bibliographic Sources*

✓ Acemoglu, D., and P. Restrepo (2017). "Robots and Jobs: Evidence from US Labor Markets." Paper presented at the annual general meeting of the American Economic Association, Chicago, January 7.

✓ Agrawal, A., J. S. Gans et A. Goldfarb (2017). " What to Expect from Artificial Intelligence ", *MIT Sloan Management Review*, vol. 58, no 3.

✓ Arthur, W. B. (2011). " The Second Economy ", *McKinsey Quarterly*, October, p. 1-9.

✓ Autor, D., D. Dorn, L. F. Katz, C. Patterson et J. Van Reenen (2017) *Concentrating on the Fall of the Labor Share*, document de travail no 23108, National Bureau of Economic Research.

✓ Baldwin, J. R., and W. Gu (2013). *Measuring Multifactor Productivity at Statistics Canada*. Statistics Canada Research Paper No. 31 in the Canadian Productivity Review Series. Catalog no. 15-206-X.

✓ Baldwin, J. R., W. Gu, R. Macdonald and B. Yan (2014). *What is productivity? How is it measured? What was Canada's productivity performance over the 1961-2012 period*," Canadian Productivity Review Research Paper No. 38, Canadian Productivity Review Series, Statistics Canada. Catalog no. 15-206-X.

✓ Bank of Sweden (2015). "Digitisation and Inflation," *Monetary Policy Report*, February, pp. 55-59.

✓ Bibbee, A. (2012). *Unleashing Business Innovation in Canada*, Working Paper No. 997, Department of Economic Affairs, Organisation for Economic Co-operation and Development.

✓ Bloom, N., R. Sadun et J. Van Reenen (2012). " Americans Do IT Better: US Multinationals and the Productivity Miracle ", *The American Economic Review*, vol. 102, no 1, p. 167-201.

✓ Bloom, N., et J. Van Reenen (2010). " Why Do Management Practices Differ Across Firms and Countries?", *Journal of Economic Perspectives*, vol. 24, no 1, p. 203-224.

✓ Boston Consulting Group (2015). *The Robotics Revolution: The Next Great Leap in Manufacturing*. Broadberry, S., B. M. S. Campbell et B. van Leeuwen (2013). " When Did Britain Industrialise? The Sectoral Distribution of the Labour Force and Labour Productivity in Britain, 1381-1851 ", *Explorations in Economic History*, vol. 50, no 1, p. 16-27.

✓ Brynjolfsson, E., and A. McAfee (2015). *Le deuxième âge de la machine: travail et prospérité à l'heure de la révolution technologique,* Paris, Odile Jacob.Cao, S., M. Salameh, M. Seki and P.

✓ St-Amant (2015). *Trends in New Firm Entry and New Entrepreneurship in Canada,* Staff Discussion Paper No. 2015-11, Bank of Canada.

✓ Cardona, M., T. Kretschmer et T. Strobel (2013). " ICT and Productivity: Conclusions from the Empirical Literature ", *Information Economics and Policy*, vol. 25, no 3, p. 109-125.

✓ Crafts, N. (2014). *Productivity Growth During the British Industrial Revolution: Revisionism Revisited*, document de travail no 204, Centre for Competitive Advantage in the Global Economy.

✓ Davis, S. J., et J. Haltiwanger (2014). *Labor Market Fluidity and Economic Performance*, document de travail no 20479, National Bureau of Economic Research.

✓ Derviş, K., et Z. Qureshi (2016). *The Productivity Slump-Fact or Fiction: The Measurement Debate*, document de travail, coll. " Global Economy and Development ", Brookings.

✓ Ericsson, N. R. (2016). *Economic Forecasting in Theory and Practice: An Interview with David F. Hendry,* Federal Reserve Board Governors, International Finance Discussion Papers No. 1184.

✓ World Economic Forum (WEF) (2016a). *Digital Transformation of Industries: Logistics Industry,* World Economic Forum white paper prepared in collaboration with Accenture.

✓ (2016b). *Shaping the Future of Construction: A Breakthrough in Mindset and Technology,* document préparé en collaboration avec le Boston Consulting Group.

✓ World Economic Forum (WEF) (2016c). *Digital Transformation of Industries: Automotive Industry,* World Economic Forum white paper prepared in collaboration with Accenture.

✓ Frey, C. B., et M. A. Osborne (2017). " The Future of Employment: How Susceptible Are Jobs to Computerisation? ", *Technological Forecasting and Social Change*, vol. 114, no C, p. 254-280.

✓ Fung, B., et H. Halaburda (2016). *Central Bank Digital Currencies: A Framework for Assessing Why and How*, document d'analyse du personnel no 2016-22, Banque du Canada.

✓ Fung, B., M. Molico and Gerald Stuber (2014). *Electronic Money and Payments: Recent Developments and Issues.* Bank of Canada Staff Working Paper No. 2014-2.

✓ Gordon, R. J. (2014a). *The Demise of U.S. Economic Growth: Restatement, Rebuttal, and Reflections*, document de travail no 19895, National Bureau of Economic Research.

✓ (2014b). *A New Method of Estimating Potential Real GDP Growth: Implications for the Labor Market and the Debt/GDP Ratio*, document de travail no 20423, National Bureau of Economic Research.

✓ (2015). " Secular Stagnation: A Supply-Side View ", *The American Economic Review,* vol. 105, no 5, p. 54-59.

✓ (2016). " Perspectives on the Rise and Fall of American Growth ", *The American Economic Review*, vol. 106, no 5, p. 72-76.

✓ Green, D. A., and B. M. Sand (2015). "Has the *Canadian* Labour Market Polarized?" *Canadian Journal of Economics.* Canadian Journal of Economics 48 (2): 612-646.

✓ Kaplan, G., B. Moll et G. L. Violante (2016). *Monetary Policy According to HANK*, document de travail no 2016/2, Council on Economic Policies.

✓ Katz, R. L., et P. Koutroumpis (2013). " Measuring Digitization: A Growth and Welfare Multiplier ", *Technovation,* vol. 33, nos 10-11, p. 314-319.

✓ Keynes, J. M. (1931). "Perspectives économiques pour nos petits- enfants", in *Essais de persuasion,* Paris, Librairie Gallimard.

✓ Krugman, P. (1997). *The Age of Diminished Expectations*, Cambridge, MIT Press.

✓ Lev, B., S. Radhakrishnan et P. C. Evans (2016). *Organizational Capital: A CEO's Guide to Measuring and Managing Enterprise Intangibles*, coll. " Measuring and Managing Organizational Capital Series ", no 1, The Center for Global Enterprise.

✓ Mendes, R. R. (2014). *The Neutral Rate of Interest in Canada.*
39

Bank of Canada Staff Working Paper No. 2014-5.

✓ Organisation for Economic Co-operation and Development (2016).

✓ *OECD Compendium of Productivity Indicators 2016,* Paris, OECD Publishing.

✓ Poloz, S. S. (2016). *From Wood Cutters to IT Specialists: Expanding Canada's Service Economy.* Speech to the C.D. Howe Institute, Toronto, November 28.

✓ Reynolds, J., et R. Cuthbertson (2014). *Retail & Wholesale: Key Sectors for the European Economy: Understanding the Role of Retailing and Wholesaling Within the European Union*, Oxford Institute of Retail Management, Saïd Business School, Université d'Oxford.

✓ Schumpeter, J. A. (1939). *Business cycles: A Theoretical, Historical, and Statistical Analysis of the Capitalist Process*, New York, McGraw-Hill Book Company.

✓ (1947). " The Creative Response in Economic History ", *Journal of Economic History*, vol. 7, no 2, p. 149-159.

✓ (1990). *Capitalisme, socialisme et démocratie,* Paris, Payot.

✓ Schwab, K. (2016). *The Fourth Industrial Revolution,* Geneva, World Economic Forum.

✓ Syverson, C. (2016). *Challenges to Mismeasurement Explanations for the U.S. Productivity Slowdown*, document de travail no 21974, National Bureau of Economic Research.

✓ Temin, P. (1997). " Two Views of the British Industrial Revolution ", *Journal of Economic History*, vol. 57, no 1, p. 63-82.

✓ Tugwell, R. G. (1931). " The Theory of Occupational Obsolescence", *Political Science Quarterly*, vol. 46, no 2, p. 171-227.

✓ Van Ark, B. (2016). " The Productivity Paradox of the New

Digital Economy ", *International Productivity Monitor*, vol. 31, p. 3-18.

✓ Van Reenen, J., N. Bloom, M. Draca, T. Kretschmer, R. Sadun, H. Overman et M. Schankerman (2010), '*The Economic Impact of ICT: Final Report'*, London, Centre for Economic Performance. Publication SMART N. 2007/0020.

✓ Varian, H. (2016). " Intelligent Technology ", *Finance and Development*, vol. 53, no 3, p. 6-9.

Table of Contents

www.ingramcontent.com/pod-product-compliance
Lightning Source LLC
Chambersburg PA
CBHW021611210326
41599CB00010B/704